YOUR KNOWLEDGE HAS VALUE

- We will publish your bachelor's and master's thesis, essays and papers

- Your own eBook and book - sold worldwide in all relevant shops

- Earn money with each sale

Upload your text at www.GRIN.com and publish for free

Prabhakar Shukla

Probiotication of beverages using whey and watermelon juice

GRIN Publishing

Bibliographic information published by the German National Library:

The German National Library lists this publication in the National Bibliography; detailed bibliographic data are available on the Internet at http://dnb.dnb.de .

Imprint:

Copyright © 2014 GRIN Verlag GmbH
Print and binding: Books on Demand GmbH, Norderstedt Germany
ISBN: 978-3-656-84387-0

PROBIOTICATION OF BEVERAGES USING WHEY AND WATERMELON JUICE

Huma Khan[1], Vocational Studies and Applied Sciences, Gautam Buddha University (GBU), Greater Noida-201308, Uttar Pradesh, India.

Hitesh Dalakoti[2], AHEC, Indian Institute of Technology Roorkee, U.K., India

ABSTRACT: *The aim of this study was to develop a probiotic beverage using whey and Watermelon juice Lactobacillus acidophilus was used as the probiotic organism. Optimization of the probiotic watermelon beverage as well as the establishment of a sensory profile by using chemical, microbiological, and quantitative descriptive analysis. For best blend optimization, we used RSM 6.0 trial software. Fermentation time using 1 per cent inoculum of L. acidophilus was optimized on the basis of RSM trial version, The 65:35 blend ratio of whey and watermelon juice fermented for 24 hrs. gave desirable results with highest sensory scores for overall acceptability and a total viable count of more than 10^6 cfu ml-1.beverage kept for 20 days below refrigerated temp and 120 hrs. at ambient temperature for the analysis for sensory and changes in titrable acidity, pH, total viable count.*

Keywords:
Whey; Probiotic beverage; watermelon whey beverage; Storage stability.

INTRODUCTION

Probiotic foods are a group of functional foods with growing market shares and large commercial interest. Probiotics are live microorganisms which when administered in adequate amounts confer a beneficial health benefit on the host. Probiotic have been used for centuries in fermented dairy products. However, the potential applications of probiotics in nondairy food products and agriculture have not received formal recognition. In recent times, there has been an increased interest to food and agricultural applications of probiotics, the selection of new probiotic strains and the development of new application has gained much importance. The uses of probiotics have been shown to turn many health benefits to the human and to play a key role in normal digestive processes and in maintaining the animal's health. The agricultural applications of probiotics with regard to animal, fish, and plants production have increased gradually. However, a number of uncertainties concerning technological, microbiological, and regulatory aspects exist. Various scientist contributed for studies of probiotic and probiotic beverages. Whey drinks are less acidic than fruit juice. The medicinal properties and nutrional properties of acidic whey can be utilized with the fruit juices. The present studies was conducted to develop a probiotic beverage using whey and watermelon juice and study of its storage stability. The co-workers are: Saavedra et al., (1994); Hosada et al., (1996) ; Majamaa and Isolauri (1996; 1997); Gionchetti et al.,(2000), Gupta et al. (2000);; De Roos and Katan, (2000);); Kyung Young Yoon et al. (2005); Sheryl H. Berman et al. (2006); K.E. Almeida et al. (2007); A.Jyothi lakshmi et al. (2011); Doherty S.B., et. al. (2011); Xiao-Hua Cui et al. (2012); M.Nor Afizah et al. (2013);Gaanappriya Mohan et al., (2013); Bathal Vijaya Kumar et al.,(2013); Priscilla Diniz Lima da Silva et al. (2013); B.N.P. Sah et al., (2013) Manasi Shukla et al., (2013); Peng Tian et al., (2014); Poonam Sharma et al., (2014); Na-Kyoung Lee et al. ,(2014).

MATERIAL AND METHOD

Toned milk procured from local market of Gurgaon. Sugar was procured from local market in packet, name Simbhaoli sugar mills. Drinking water supplied by fare lab was used in the preparation whenever needed. Watermelon was procured from local market of Gurgaon, and fresh juice was prepared in lab. The freeze dried culture Lactobacillus acidophilus, NDCC-015 the lyophilized culture used as probiotic in beverage, was procured from NDRI Karnal (Haryana).

COLLECTION OF MILK SAMPLE & SEPARATION OF WHEY

Toned milk was heated in a stainless steel vessel to 80°C followed by cooling to 70°C. The hot milk was acidified by adding 2 per cent citric acid solution followed by continuous stirring which resulted in the complete coagulation of the milk protein (casein). The liquid (whey) was filtered using muslin cloth. The prepared whey was heated to 85°C before blending with fruit juice and sugar was added directly to whey when it was heated to 85 °C before mixing. Watermelon were de-crowned and peeled. The fruit slices were grinded and the resulting juice was passed through a double layered muslin cloth to extract the clear juice. The freeze dried culture was activated using M.R.S (de Man, Rogosa and Sharpe) Broth.

FORMULATION OF DEVELOPED BEVERAGE

The optimization of beverage was performed through RSM with central composite design (CCD) using Design-Expert 6.0 (trial version) software. Three factors; watermelon juice, stabilizer and sucrose were selected as a process parameter (independent) while consistency, colour and appearance, flavor and overall acceptability as response parameter (dependent). The concentration of watermelon juice, stabilizer and sucrose in beverage formulation were optimized by central composite design (CCD). These ingredient influence the consistency, colour and appearance, Flavor and overall acceptability of beverage up to major extent and their ranges were selected in accordance with literature and preliminary experiments carried out in the laboratory.

INOCULUMS PREPARATION

For preparation of inoculums, cells from stock culture were transferred to 25 mL sterile de Man, Rogosa and Sharpe (**MRS**) broth and incubated for 48 h at 37 °C. The growth density was measured at a wavelength of 620 nm using a spectrophotometer. The bacteria were kept under stationary conditions until they reached an absorbance of 5.0. One ml of inoculums contained around 10^9 colony forming units (CFU). Dilutions were made with MRS broth to bring the lactobacillus to an optical density at 620 nm of 0.1 ± 0.02 corresponding to a cell density of 10^5–10^7 CFU/mL.

PROBIOTIC BEVERAGE PREPARATION

For the production of the fermented beverage, best blend selected using RSM was prepared by reconstituted whey, watermelon juice, sugar and pectin. The blend was selected for acidification with Lactobacillus acidophilus.

FERMENTATION

The incubation period was optimized by inoculating blend .The beverage was evaluated for sensory characteristic, total viable count, pH and titrable acidity for sample fermented for 5, 10, 15, 20 and 24 h .

STORAGE

The prepared sample was stored in two conditions. In first condition the sample were stored at room temperature while in second condition the sample were stored at refrigeration below 10°C.Change in sensory characteristic, total viable count, pH and titratable acidity were studied during storage and sample were stored at refrigeration below 10°C.Change in sensory characteristic, total viable count, pH and titratable acidity were studied during storage.

ANALYTICAL METHOD

The total soluble solid (Brix) was determined was determined by using a calibrated hand refractometer (ERMA) range is $0\text{-}32^0B$. The pH of the beverage was determined using a digital pH meter. Titratable acidity was determined according to the AOAC method .Sugars is estimated by Lane and Eynon method .Ash and Moisture estimated according to AOAC.

STATISTICAL ANALYSIS

Response surface methodology (RSM) was adopted in the experimental design as it emphasizes the modelling and analysis of the problem in which response of interest is influenced by several variables and the objective is to optimize this response. The main advantage of RSM is reduced number of experimental runs needed to provide sufficient information for statistically acceptable results. A five-level three-factor central composite rotatable design was employed. The independent variables were the watermelon juice(X1), sugar (X2), and stabilizer (X3). The variables and their levels were chosen based on the limited literature available different blend of fruits. These were the watermelon juice (X1; 10-35), sugar (X2; 8-14), and stabilizer (X3; 0.10-0.30). The experimental design matrix in coded (x) form and at the actual level (X) of variables is given in Table 1, A total of 20 experiments were carried out by using watermelon juice under different experimental conditions as given in Table 3.

Table 1: Coded and Non-Coded Values of Variables and Their Levels for Preparation of Beverage

Independent Variable	Non-Coded Value	Coded Level				
		-α	-1	0	1	+α
Watermelon Juice	X1	1.47759	10	22.5	35	43.52241
Sugar	X2	5.954622	14	11	8	16.04538
Pectin	X3	0.031821	0.1	0.2	0.3	0.368179

(Source: Own Data)

ANALYSIS OF DATA: The second order polynomial response models were fitted for each response variables (Y) with the independent variable (x):

$$Y = \beta_0 + \sum_{i=1}^{k} \beta_i x_i + \sum_{i=1}^{k} \beta_{ii} x_i^2 + \sum_{i=1}^{k-1} \sum_{j=i+1}^{k} \beta_{ij} x_i x_j$$

Where, Y is the response, k = 3, $\beta 0$ is the constant coefficient, xi (i = 1-3) are non-coded variables, βi are linear, βii are the quadratic and βij (i and j = 1-3) are the second order interaction coefficients. The mathematical models were evaluated for each response by means of multi linear regression analysis. The significant terms in the model were found by Analysis of Variance (ANOVA) for each response with 95% of confidence. Significance was judged by determining the probability level that the F-statistic calculated from the data is less than 5%. The adequacy of the model was determined using model analysis, lack of fit test, coefficient of determination R^2 (Myers and Montgomery, 1995). The model generally considered adequate when the calculated F-value is more than significant value. Maximization and minimization of the polynomials thus fitted was performed by desirability function method and mapping of the fitted response was achieved using Design-Expert 6.0 (trial version) software.

RESULT AND DISCUSSION

The prepared beverage (65w; 35w) had Acidity, total sugar, Ash, Moisture 0.11, 5.8, 0.29, 91.0 respectively.

EXPERIMENTAL DESIGN-
The main advantages of RSM is the reduce number of experimental runs needed to provide sufficient information for statistically acceptable result. Three variables (five level of each variable), central composite rotable experimental design (CCRD) was employed. The independent variables considered were juice composition(x_1), stabilizer(x_2) and sucrose(x_3). The five level of the process variables were coded as -2,-1,0,+1,+2 (Montgomesy 2001). To examine the combined effect of the three independent variables on preparation of juice used, a face centered cable design of $2^3=8$ plus 6 center points and $(2\times3=6)$ .star points leading to a total of 20 experiments (Table 4.4) were performed.

Table 2: Experimental Design in Coded Form for Responsible For Response Surface Analysis. (Source: Own Data)

Coded variables			Combinations	Replications	No. of Experiments
X_1	X_2	X_3			
±1	±1	±1	8	1	8
±2	0	0	2	1	2
0	±2	0	2	1	2
0	0	±2	2	1	2
0	0	0	1	6	6

Table 3: Experimental Central Composite Design CCD Runs In Design Expert (6.0) Trial Version Software And Corresponding Results.

S.NO	Watermelon juice%	Sucrose%	Stabilizer%	Consistency	Colour and appearance	Flavor	Overall acceptability
1	1.4	11	0.2	7.68	7.1	7.21	7.21
2	22.5	11	0.2	8.12	8.62	8.8	8.58
3	10	14	0.3	7.21	7.32	7.35	7.55
4	22.5	11	0.36	7.11	8.12	7.22	7.68
5	22.5	16	0.2	7.54	7.12	7.78	7.22
6	10	8	0.1	6.9	7.1	6.67	6.52
7	22.5	11	0.2	8.23	8.55	8.67	8.67
8	22.5	11	0.2	8.2	8.35	8.55	8.45
9	22.5	11	0.2	8.31	8.3	8.46	8.56
10	35	8	0.1	7.12	7.8	7.92	7.66
11	10	8	0.3	7.22	7.65	7.32	7.44
12	10	14	0.1	7.43	7.11	7.45	7.22
13	35	14	0.3	7.92	7.22	7.33	7.32
14	22.5	11	0.03	7.12	8.12	8.45	8.1
15	35	14	0.1	7.44	7.56	7.34	7.22
16	35	8	0.3	7.34	7.44	7.45	7.22
17	22.5	11	0.2	8.22	8.68	8.55	8.63
18	22.5	11	0.2	8.34	8.55	8.43	8.44
19	22.5	5.95	0.2	7.22	7.35	7.43	7.56
20	43.52	11	0.2	7.72	7.65	7.34	7.45

(Source: Own Data)

ADEQUACY OF MODELS FOR DIFFERENT REFERENCES

The ANOVA results of the second order polynomial response model in Table 3 indicated that the model equation derived by Design-Expert 6.0 (trial version) software could adequately be used to describe the influence of response parameter on process parameters under the desired operating conditions. The estimated regression coefficients of the quadratic polynomial models for the response variables, along with the corresponding R^2, adj- R^2 , Pred-R^2, are shown in below table .Analysis of variance showed that the models were highly significant (p<0.001) for all the responses. The lack of fit which measures the fitness of the models did not result in a significant F-value in all response parameters, indicating that these models were sufficiently accurate for predicting those responses. The coefficient of determination (R^2) values of all responses are quite high and approaching to unity, indicating a high proportion of variability was explained by the data and RSM models were adequate. The Coefficient of Variation (CV) should not be greater than 10% but in this study, it was found in the range of 1.9- 3.2% for all the responses, which indicates better precision and reliability of the experiments carried out.

Table 4: Anova Results Showing the Variables as a Linear, Quadratic and Interaction Terms on Analysis Of Variance for Consistency of Product

Factor	Coefficient estimate	Sum of Squares	Standard error	df	F-value	Prob>F
A	.083	0.093	0.040	1	4.25	.0663
B	0.14	0.28	0.040	1	12.81	0.0050
C	0.057	0.045	0.040	1	2.05	0.1828
AB	0.048	0.018	0.052	1	0.082	0.3855
AC	0.075	0.045	0.052	1	0.45	0.5188
BC	-0.035	-003	0.052	1	0.45	0.3855
A^2	-0.20	0.55	0.039	1	25.01	0.0005
B^2	-0.31	1.37	0.039	1	62.44	<0.0001
C^2	-0.40	2.33	0.039	1	106.19	<0.0001

Table 5: Analysis Of Variance Result of Consistency

Response	Source	Sum of squares	df	Mean squares	F-value	P-value
LE	Residual	0.22	10	0.022		
	Lack of Fit	0.19	5	0.038	5.99	0.0357
	Pure Error	0.031	5	-003		
	Cor Total	4.34	19			

Table 6: Analysis Of Variance for Colour 6 and Appearance of Beverage

Factor	Coefficient estimate	Sum of Squares	Standard error	df	F-value	Prob>F
A	0.13	0.23	0.037	1	11.97	0.0061
B	-0.085	0.100	0.037	1	5.23	0.0452
C	-0.03	-0.04	0.037	1	0.014	0.9087
AB	-0.017	-0.03	0.049	1	0.13	0.7274
AC	-0.18	0.27	0.049	1	13.98	0.0038
BC	-0.040	0.013	0.049	1	0.67	0.4315
A^2	-0.43	2.62	0.036	1	137.29	<0.0001
B^2	-0.48	3.26	0.036	1	171.05	<0.0001
C^2	-0.16	0.38	0.036	1	20.01	0.0012

(Source: Own Data)

Table 7: Analysis Of Variance Result of Colour and Appearance

Response	Source	Sum of squares	df	Mean squares	F-value	P-value
LE	Residual	0.19	10	0.019		
	Lack of Fit	0.077	5	0.015	0.67	0.6627
	Pure Error	0.11	5	0.023		
	Cor Total	6.24	19			

(Source: Own Data)

Table 8: Analysis Of Variance for Flavor of Beverage

Factor	Coefficient estimate	Sum of Squares	Standard error	DF	F-value	Prob>F
A	0.11	0.16	0.081	1	1.77	0.2134
B	0.051	0.036	0.081	1	0.40	0.5415
C	-0.15	0.29	0.081	1	3.27	0.1006
AB	-0.19	0.29	0.11	1	3.19	0.1045
AC	-0.13	0.13	0.11	1	1.48	0.2513
BC	-0.036	0.011	0.11	1	0.12	0.7388
A^2	-0.49	3.51	0.079	1	39.22	< 0.0001
B^2	-0.38	2.04	0.079	1	22.86	0.0007
C^2	-0.30	1.26	0.079	1	14.06	0.0038

(Source: Own Data)

Table 9: Analysis Of Variance Result of Flavor

Response	Source	Sum of squares	Df	Mean squares	F-value	P-value
LE	Lack of Fit	0.80	5	0.16	8.40	0.0178
	Pure Error	0.095	5	0.019		
	Cor Total	7.57	19			

(Source: Own Data)

Table 10: Analysis Of Variance for Overall Acceptability of Beverage

Factor	Coefficient estimate	Sum of Squares	Standard error	DF	F-value	Prob>F
A	0.080	0.088	0.067	1	1.43	0.2596
B	-003	-004	0.067	1	0.012	0.9136
C	0.015	-003	0.067	1	0.050	0.8284
AB	-0.14	0.17	0.088	1	2.70	0.1316
AC	-0.20	0.32	0.088	1	5.15	0.0466
BC	-003	-004	0.088	1	-003	0.9445
A^2	-0.48	3.27	0.065	1	53.39	< 0.0001
B^2	-0.46	2.99	0.065	1	48.74	< 0.0001
C^2	-0.28	1.12	0.065	1	18.25	0.0016

(Source: Own Data)

Table 11: Analysis Of Variance Result of Overall Acceptability

Response	Source	Sum of squares	df	Mean squares	F-value	P-value
LE	Residual	0.61	10	0.061		
	Lack of Fit	0.57	5	0.11	13.01	0.0068
	Pure Error	0.044	5	-0.03		
	Cor Total	7.44	19			

Table 12: Optimized Levels and Predicted Optimum Value of Response

%	Optimum value	Responses	Predicted value
Watermelon juice%	24.19657	Consistency	8.244359
		Color and appearance	8.519445
		Flavor	8.596069
Sugar%	11.09258	Overall acceptability	8.562279
Stabilizer%	0.194334	Desirability	0.92544

RESPONSE SURFACES DEVELOPED TO OBSERVE THE EFFECT OF PROCESS PARAMETERS ON RESPONSE PARAMETERS

In the beverage production process watermelon, stabilizer and sucrose are taken process parameters while consistency, appearance, flavor and overall acceptability are taken as response parameters.

Table 13: Effect of Lactobacillus Acidophilus On Sensory Characteristics of Whey and Watermelon Juice at $37 \pm 1°C$.

Fermentation time(h)	Consistency	Color and appearance	Flavor	Overall acceptability
5	8.55	8.64	8.63	8.72
10	8.23	8.29	8.21	8.4
15	8.11	8.03	8.03	7.69
20	7.63	7.57	7.62	6.70
24	6.66	6.63	6.04	5.22

(Source: Own Data)

Table 14: Effect of Fermentation Period On The pH, Titratable Acidity and Total Viable Count of Whey Blend at Ambient Temperature (Source: Own Data)

Fermentation time(h)	pH	Blend	
		Titrable acidity%	Total viable count
5	4.3	0.546	3.80×10^7
10	4.3	0.605	4.92×10^7
15	4.2	0.806	6.22×10^8
20	3.7	0.896	7.90×10^8
24	3.8	0.926	8.80×10^8

Table 15: Changes in Sensory Characteristic of Probiotic Beverage during Storage for 0-20 Days. (Source: Own Data)

(SP) at 5°c	Consistency	C&A	Flavor	OAA
0	8.66	8.78	8.40	8.59
5	8.70	8.82	8.30	8.34
10	8.67	8.64	8.44	8.54
15	8.73	8.68	8.32	8.65
20	7.97	8.26	8.21	7.88

Table 16: Changes in Sensory Characteristic of Probiotic Beverage during Storage for 0-120 Hours (Source: Own Data)

SP(h) at 37°C	Consistency	C &A	flavor	OAA
0	8.65	8.11	8.54	8.45
24	8.05	7.92	8.02	8.22
48	7.05	7.25	7.34	7.20
72	6.34	6.20	6.67	6.54
96	4.60	4.56	4.65	4.20
120	4.22	4.12	3.50	3.81

Table 17: Changes in pH, Total Viable Count and Titratable Acidity of Probiotic Beverage during Storage For 0-20 Days.   (Source: Own Data)

(SP) at 5°c	pH	Total viable count(Cfu/ml)	Titrable acidity
0	4.36	3.8×10^7	0.546
5	4.36	3.0×10^7	0.56
10	4.35	4.2×10^7	0.58
15	4.32	3.1×10^7	0.59
20	4.32	2.0×10^7	0.60

Table 18: Changes in pH, Total Viable Count and Titratable Acidity of Probiotic Beverage during Storage for 0-120 Hours. (Source: Own Data)

SP(h) at 37°c	pH	Total viable count	Titrable acidity
0	4.3	3.8×10^7	0.546
24	4.3	8.0×10^7	0.59
48	4.2	8.5×10^7	0.77
72	4.1	6.2×10^7	0.79
96	4.1	3.3×10^7	0.80
120	3.9	3.0×10^7	0.83

OPTIMIZATION OF GROWTH CONDITIONS FOR LACTOBACILLUS ACIDOPHILUS IN PROBIOTIC BEVERAGE DEVELOPMENT

Variations in fermentation time and medium were studied in terms of overall acceptability of the beverage, total viable counts, pH and titratable acidity. Whey supplemented with 11% sucrose was fermented for different time intervals with the addition of watermelon juice. The effect of Lactobacillus acidophilus on sensory attributes of whey with watermelon juice and pectin during fermentation has been shown above. Color and appearance of the beverage was significantly affected by the incubation period. The mean scores for color and appearance ranged from 8.92 to 6.53 for whey with beverage. The mean score was highest (8.92) for whey fermented beverage for 5 h. The sensory scores for consistency reduced significantly with increase in fermentation time for whey-watermelon juice blend. The sensory score for consistency of fermented whey ranged from 8.55 to 6.66. The sensory scores for consistency of fermented whey watermelon juice blend ranged from 8.69 to 6.04. The sensory scores for color and appearance ranged from 8.64 to 6.63 for fermented whey-watermelon juice blend. The mean score for flavor decreased significantly with increasing fermentation time irrespective of the medium.

The mean score for overall acceptability ranged from 8.72 to 5.22 of whey-watermelon juice blend. Highest score for overall acceptability was seen in case of whey-watermelon juice blend fermented for 5 h.

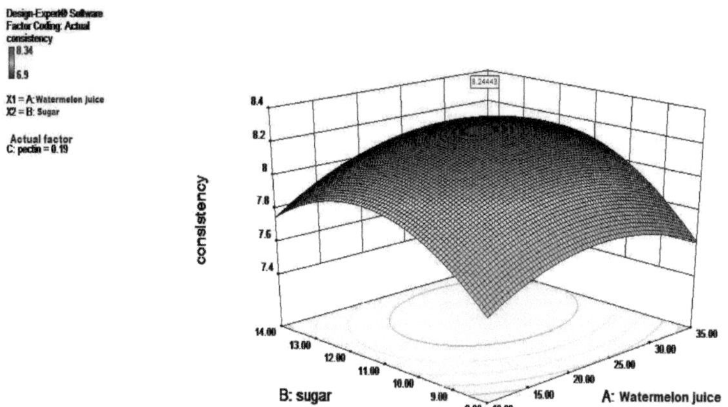

Fig 1: Response surface for the effect of watermelon and sugar on the consistency of beverage. (Source: Own Data)

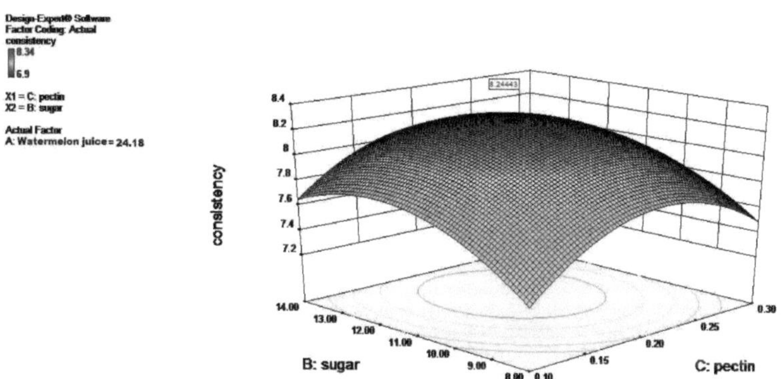

Fig 2: Response surface for the effect of Pectin and sugar on the consistency of beverage. (Source: Own Data)

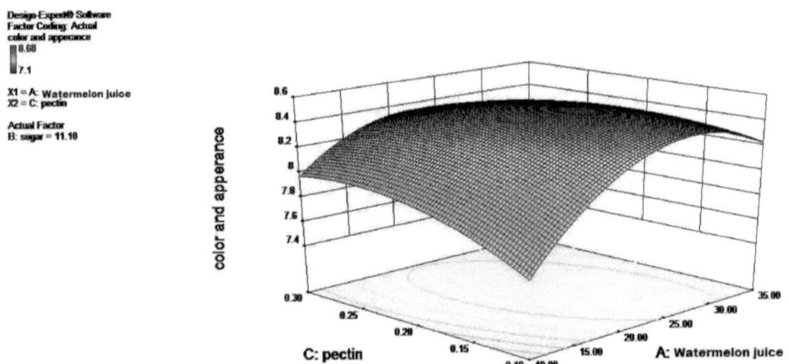

Fig 3: Response surface for the effect of Pectin and juice on color and appearance of beverage. (Source: Own Data)

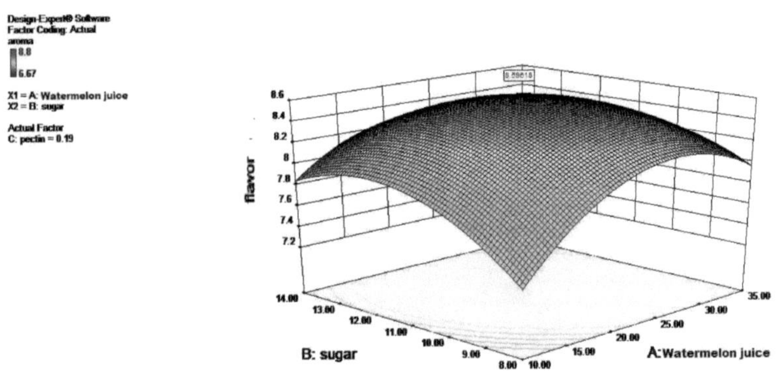

Fig 4: Response surface for the effect of sugar and watermelon juice on flavor of beverage.

(Source: Own Data)

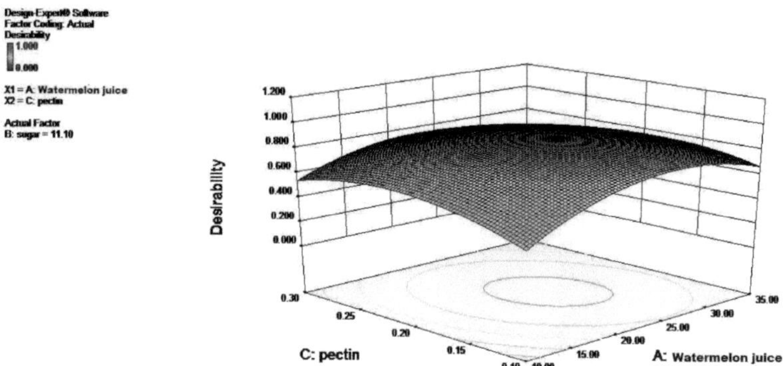

Fig 5: Response surface for the effect of Pectin and juice on Desirability of beverage
(Source: Own Data)

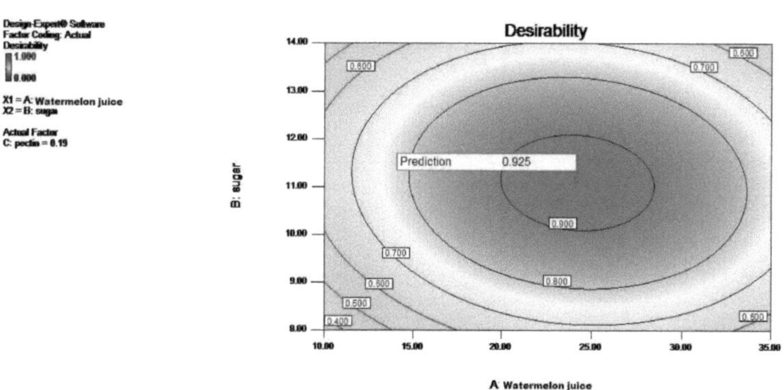

Fig 6: Response surface for the effect of Sugar and Watermelon juice Desirability of
beverage. (Source: Own Data)

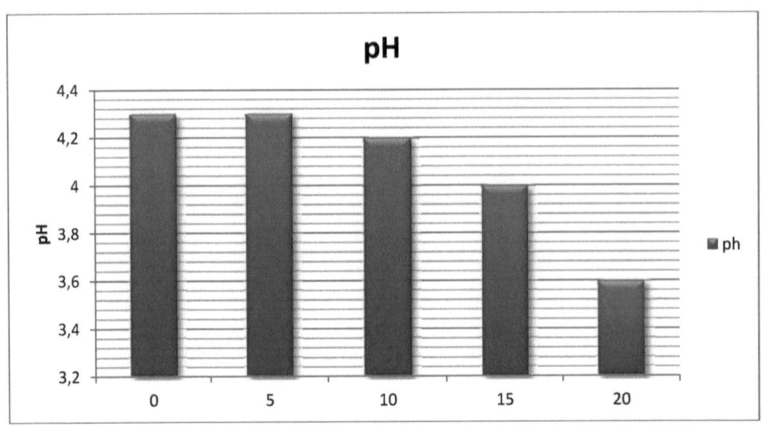

Fig 7: Change in pH during fermentation at incubation temp (appox.37°C).
(Source: Own Data)

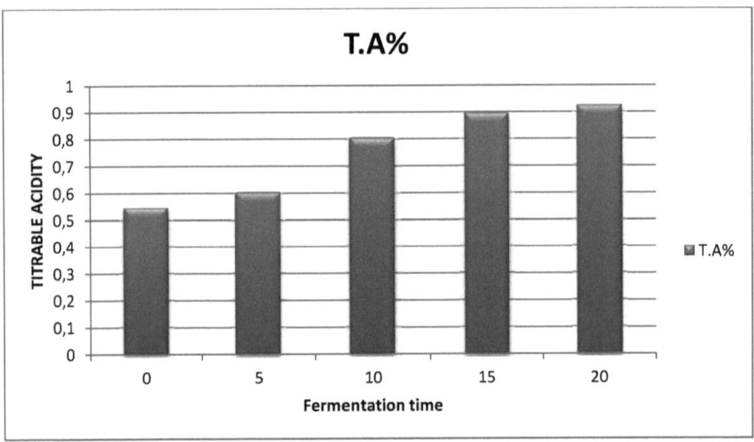

Fig 8: Change in titratable acidity during fermentation (approx. 37°C). (Source: Own Data)

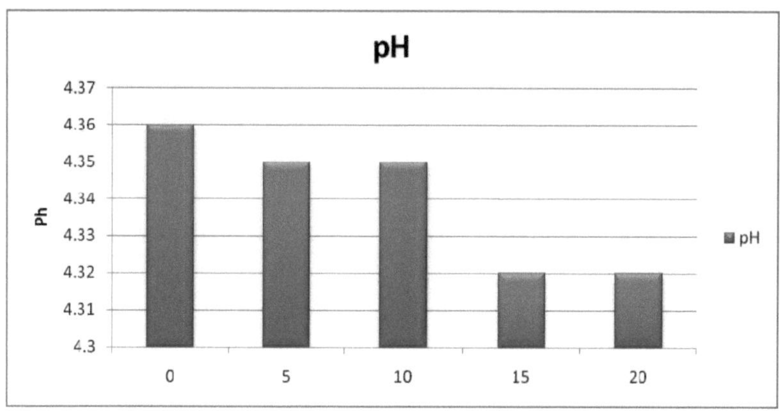

Fig 9: Change in pH during storage for 20 days at refrigerated temperature (below 10ºC)
(Source: Own Data)

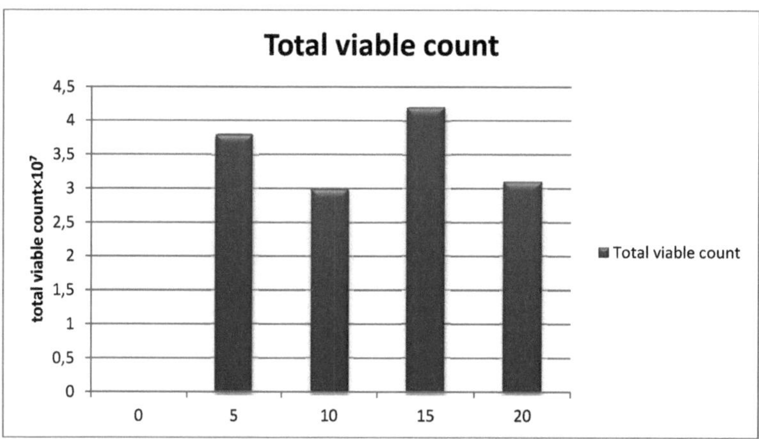

Fig 10: Change in total viable count during storage at refrigerated temp (below 10ºC).
(Source: Own Data)

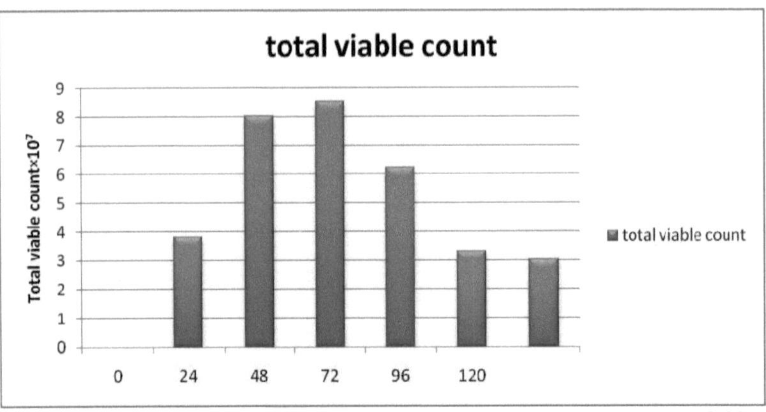

Fig 11: Change in pH during storage at ambient temperature (appox.37°C)
(Source: Own Data)

CONCLUSION

Response Surface Methodology (RSM) was applied successfully for the optimization of beverage formulation for production of instant energy probiotic beverage, using whey as base material. The watermelon juice, stabilizer and sucrose were taken as process parameters and consistency, appearance, taste, aroma and overall acceptability were taken as response parameters. The effect of watermelon juice, stabilizer and sucrose on response parameters were also analyzed through Design-Expert 6.0 (trial version) software. The second order polynomial response models were fitted for each response variables, RSM and ANOVA tests were performed. The correlation coefficients for all the response models were found to be quite high (>0.9936). Optimum values for watermelon juice, stabilizer and sucrose were found to be 24.18, 0.19 and 11.0%, respectively. The prepared beverage was highly palatable and nutritionally.

REFERECES

1. **Jyothi lakshmi, Purnima Kaul(2011)** Nutritional potential, bioaccessibility of minerals and functionality of watermelon (Citrullus vulgaris) seeds,. *Food Science and Technology* **44** :1821-1826.

2. **B.N.P. Sah, T. Vasiljevic , S. McKechnie, O.N. Donkor (2013).** Effect of probiotics on antioxidant and antimutagenic activities of crude peptide extract from yogurt. *Food Chemistry* 156: 264–270.

3. **Bathal Vijaya Kumar, Mannepula Sreedharamurthy, Obulam Vijaya Sarathi Reddy(2013).**Physico-chemical analysis of fresh and probioticated fruitjuices with *lactobacillus case. Int J Appl Sci Biotechnol,* Vol. **1(3)**: 127-131.

4. **De Roos, N. M., & Katan, M. B. (2000).** Effects of probiotic bacteria on diarrhea, lipid metabolism, and carcinogenesis: a review of papers published between 1988 and 1998. *The American journal of clinical nutrition, 71*(2), 405-411.

5. **Doherty S.B (2011).** Development and characterization of whey protein micro-beads as potential matrices for probiotic protection. Food Hydrocolloids 25:1604-1617.

6. **Gaanappriya Mohan, Guhankumar P., KiruththicaV., Santhiya N. and Anita S.(2013)** Probiotication of fruit juices by lactobacillus acidophilus. Vol **4**, Issue 1, 2013, pp 72-77.

7. **Gionchetti, P., Rizzello, F., Venturi*, A., Brigidi†, P., Matteuzzi, D., Bazzocchi, G., Poggioli , G., Miglioli*, M., & Campieri*, M. (2000).** Oral bacteriotherapy as maintenance treatment in patients with chronic pouchitis: a double-blind, placebo-controlled trial.*Gastroenterology, 119*(2), 305-309.

8. **Gupta, P., Andrew, H., Kirschner, B. S., & Guandalini, S. (2000).** Is Lactobacillus GG helpful in children with Crohn's disease? Results of a preliminary, open-label study.*Journal of Pediatric Gastroenterology and Nutrition, 31*(4), 453-457.

9. **K.E. Almeida , A.Y. Tamime , M.N. Oliveira(2007).** Influence of total solids contents of milk whey on the acidifying profile and viability of various lactic acid bacteria LWT - *Food Science and Technology* **42** : 672–678.

10. **Kyung Young Yoon, Edward E. Woodams, Yong D. Hang(2005).** Production of probiotic cabbage juice by lactic acid bacteria.*Bioresource Technology* **97** :1427–1430.

11. **M. Nor Afizah , Syed S.H. Rizvi(2013).**Functional properties of whey protein concentrate texturized at acidic pH: Effect of extrusion temperature. *Food Science and Technology* **57**:290-298.

12. **Majamaa, H., Isolauri , E., Saxelin, M., & Vesikari, T. (1995).** Lactic acid bacteria in the treatment of acute rotavirus gastroenteritis. *Journal of Pediatric Gastroenterology and Nutrition, 20*(**3**), 333-338.

13. **Manasi Shukla, Yogesh Kumar Jha1, and Shemelis Admassu(2013).** Development of Probiotic Beverage from Whey and Pineapple Juice. *J Food Process Technol 4: 206.*

14. **Na-Kyoung Lee, So-Yeon Kim, Kyoung Jun Han, Su Jin Eom, Hyun-Dong Paik(2014)** Probiotic potential of Lactobacillus strains with anti-allergic effects from kimchi for yogurt starters. *Food Science and Technology* **58**:130-134.

15. **Poonam Sharma, Sudhir Kumar Tomar, Pawas Goswami, Vikas Sangwan, Rameshwar Singh (2014).**Antibiotic resistance among commercially available probioticsFood Research International **57** (2014): 176–195

16. **Priscilla Diniz Lima da Silva, Maria de Fátima Bezerra, Karina Maria Olbrich dos Santos, Roberta Targino Pinto Correia (2013).** Potentially probiotic ice cream from goat's milk: Characterization and cell viability during processing, storage and simulated gastrointestinal conditionsLWT - *Food Science and Technology* **xxx** : 1-6.

17. **Saavedra, J. M., Bauman, N., Perman, J., Yolken, R., & Oung, I. (1994).** Feeding of Bifidobacterium bifidum and Streptococcus thermophilus to infants in hospital for prevention of diarrhoea and shedding of rotavirus. *The Lancet, 344*(**8929**), 1046-1049.

18. **Xiao-Hua Cui, Shu-Jun Chen, Yu Wang, Jian-Rong Han(2012).** Fermentation conditions of walnut milk beverage inoculated with kefir grains. Food Science and Technology 50:349-352.